Gene Therapy

Mary C. Colavito
Santa Monica College

Michael A. Palladino, Series Editor
Monmouth University

San Francisco   Boston   New York
Cape Town   Hong Kong   London   Madrid   Mexico City
Montreal   Munich   Paris   Singapore   Sydney   Tokyo   Toronto

Acquisitions Editor: Susan Winslow
Editorial Assistant: Mercedes Grandin
Marketing Manager: Lauren Harp
Managing Editor: Mike Early
Production Supervisor: Lori Newman
Production Management: Black Dot Group
Composition and Illustration: Black Dot Group
Manufacturing Buyer: Stacy Wong
Text Designer: Black Dot Group
Photo Research: Travis Amos
Director, Image Resource Center: Melinda Patelli
Image Rights and Permissions Manager: Zina Arabia
Cover Designer: Seventeenth Street Studios
Cover Illustration: Seventeenth Street Studios
Cover/Text Printer: Courier/Stoughton

ISBN 0-8053-3819-5

Copyright ©2007 Pearson Education, Inc., publishing as Benjamin Cummings, 1301 Sansome St., San Francisco, CA 94111. All rights reserved. Manufactured in the United States of America. This publication is protected by Copyright and permission should be obtained from the publisher prior to any prohibited reproduction, storage in a retrieval system, or transmission in any form or by any means, electronic, mechanical, photocopying, recording, or likewise. To obtain permission(s) to use material from this work, please submit a written request to Pearson Education, Inc., Permissions Department, 1900 E. Lake Ave., Glenview, IL 60025. For information regarding permissions, call (847) 486-2635.

Many of the designations used by manufacturers and sellers to distinguish their products are claimed as trademarks. Where those designations appear in this book, and the publisher was aware of a trademark claim, the designations have been printed in initial caps or all caps.

1 2 3 4 5 6 7 8 9 10—CRS—10 09 08 07 06

www.aw-bc.com

Contents

Realizing the Promise of Gene Therapy: Successes, Setbacks, and Challenges   1

Gene Therapy Methods   3

Identifying a Disease That Is Likely to Respond to Gene Therapy   3

Isolating a Functional Copy of the Gene   8

Incorporating the Gene into a Carrier for Gene Delivery   10

Determining Whether the Gene Product Is Made   15

Partial Successes and Ongoing Trials   16

Gene Therapy for Severe Combined Immune Deficiency (ADA-SCID)   16

Gene Therapy for Cystic Fibrosis   17

Gene Therapy for Canavan Disease   18

Setbacks   20

Immune Reactions Targeting Cells Carrying the Vector   20

Inadvertent Activation of Cancer-Causing Genes by a Vector   20

Insufficient Numbers of Cells Producing the Gene Product   22

Challenges 22
- *Safe and Effective Delivery of Genes 23*
- *Production of a Sufficient Amount of Gene Product 30*
- *Achieving a Lasting Improvement for the Patient 31*

Future Approaches 31
- *Correction of a Patient's DNA 31*
- *Interfering with the Production of the Gene Product 33*
- *Treating Complex Conditions Resulting from Multiple Gene Defects 38*

Conclusion 39

Resources for Students and Educators 39
- *For Students 39*
- *For Educators 41*

Realizing the Promise of Gene Therapy: Successes, Setbacks, and Challenges

The tiny patient's recovery was dependent upon the delivery of a novel medication, one that would program his own cells to continually produce the protein he needed. Before his birth, Andrew Gobea, shown with his physician in Figure 1, was diagnosed with Severe Combined Immune Deficiency (**ADA-SCID**) due to the absence of an enzyme called Adenosine Deaminase (ADA). Insufficient levels of this enzyme would rob him of the disease-fighting capabilities of a healthy immune system. Andrew inherited this disease from his healthy parents, each of whom carry one functional and one defective copy of the gene giving instructions for producing ADA. Andrew's cells are unable to produce their own ADA because he inherited both copies of the defective *ADA* genes. In anticipation of his birth, Dr. Donald Kohn's research team at Children's Hospital in Los Angeles began preparing a gene therapy treatment that would deliver functional copies of the *ADA* gene to Andrew's immune system cells. The goal was to successfully introduce the working *ADA* genes into cells that would remain in Andrew's bone marrow, dividing throughout his lifetime to give rise to white blood cells capable of fighting infections. Andrew was a sleepy, four-day old infant when he received his dose of potentially life-saving genes. He would be kept alive on a costly protein replacement therapy involving repeated injections of ADA while his doctors waited to see whether ADA-producing cells survived and multiplied in his system.

While this scenario may seem highly futuristic, gene therapy trials have actually been carried out for the past 15 years. The goal of gene therapy can be stated simply: to provide a patient with a functional gene to counteract the effects of a non-functional version. Yet the procedure is complex, requiring the optimization of many processes, involving methods to deliver the genes to a sufficient number of cells of a specific type. All gene therapy treatments are still experimental, at various stages in the process toward approval for regular use in a clinical setting. The progress toward effective gene therapy has been uneven, leading to partial successes in some cases and significant setbacks in others. Yet the tantalizing promise of permanent cures for fatal diseases spurs physicians and researchers to find better ways of meeting the

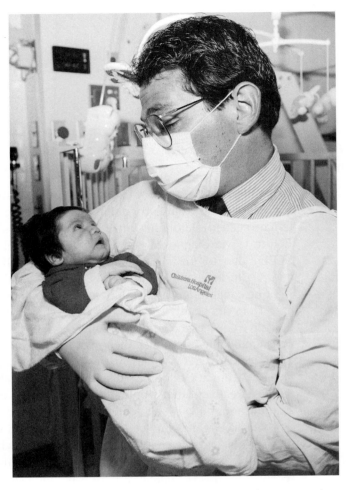

FIGURE 1. In 1993, Andrew Gobea was one of the first infants to receive gene therapy treatment for ADA-SCID under the care of Dr. Donald Kohn.
Source: AP Photo

challenges of effective gene delivery to help the unfortunate people who have been betrayed by their own genes. The purpose of this booklet is to provide you with an introduction to gene therapy including discussions of the methodology along with the progress and pitfalls of the procedures. You can follow the history of gene therapy successes and setbacks using the timeline in Table 1. Throughout this booklet, key terms appear in bold to help you learn important concepts related to gene therapy.

TABLE 1. Timeline for Gene Therapy Successes and Setbacks

Year	Event
1990	Ashanthi DeSilva and one other child treated for ADA-SCID.
1993	Andrew Gobea and two other infants treated for ADA-SCID.
1997	Liposome delivery of genes to treat cystic fibrosis patients in Scotland.
1997	Gene therapy with tumor suppressor gene p53 causes regression of lung cancer tumors.
1999	Jesse Gelsinger is treated for ornithine transcarbamylase (OTC) deficiency and dies as a result of a massive immune reaction.
2000	Successful X-SCID treatment for nine patients in France.
2000	Gene therapy for Hemophilia B, Factor IX deficiency, provides a partial benefit in two patients.
2000	Successful gene therapy treatment of head and neck cancers.
2001	Treatment of Canavan disease with a viral vector.
2002	Successful ADA-SCID treatment for two patients in Italy.
2002	Report of leukemia development in three X-SCID patients in France. Gene therapy trial put on hold.
2004	Death of one X-SCID gene therapy recipient in France.
2005	Gene therapy slows progression of Alzheimer's disease.
2005	Improvement of motor skills for gene therapy treatment of Parkinson's Disease.
2006	Report of 18-month correction of chronic granulomatous disease in two patients.

Gene Therapy Methods

IDENTIFYING A DISEASE THAT IS LIKELY TO RESPOND TO GENE THERAPY

How do scientists select a disease that is likely to respond to gene therapy? There have been three main criteria so far. First, the disease must be the result of a defect in a single gene. Secondly, this gene must be identified and purified in its functional form. Third, the cells affected by the gene defect must be accessible for gene delivery. Many of the diseases chosen for gene therapy are those for which there are few, if any, other effective treatments.

Early attempts have focused on blood cells, airway cells, muscle cells, and brain cells as recipients of functional genes. Gene therapy has been restricted to these and other **somatic cells**, which do not give rise to sperm or egg cells. The benefit is intended to treat cells and tissues most affected by the disease but not to be passed along to the next generation through the germline or sex cells.

Although the precise number is not yet known, the most recent estimates predict that there are 25,000–35,000 genes in each human cell. Genes are composed of the hereditary material DNA, and each gene represents a unique arrangement of DNA nucleotide building blocks. While some genes are used by nearly every cell to maintain common functions involved in metabolism, other genes are active only in certain cells and give them unique characteristics. The instructions from an active gene allow the cell to make a specific protein. As shown in Figure 2, these instructions are first copied into RNA, a genetic intermediate, by a process called transcription. The RNA copy is then used to synthesize a protein.

Proteins work together to allow the cell to carry out specific functions. For example, the *ADA* gene allows the enzymatic protein ADA to be produced. This enzyme participates in a biochemical pathway that produces the components necessary to duplicate DNA prior to cell division and to repair DNA if any damage has occurred. Cells of the immune system are particularly sensitive to decreased levels of ADA because DNA duplication and repair are essential to their disease-fighting activities. Immune system cells divide rapidly in response to bacterial and viral invaders, so the inability to duplicate

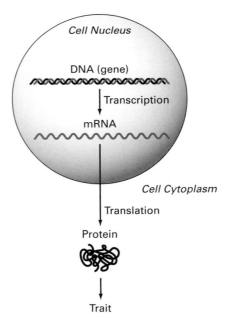

FIGURE 2. The flow of genetic information in human cells.
Source: Palladino, *Understanding the Human Genome Project* 2e, Pearson Benjamin Cummings, Figure 3

DNA hinders recovery from an infection. In addition, some of these cells undergo gene rearrangements in order to prepare them to produce specific disease-fighting proteins. The rearrangement process generates breaks in the DNA that, if not repaired, trigger cell destruction. The block in DNA synthesis also leads to a buildup of metabolites that are toxic to immune system cells. Therefore, defects in the *ADA* gene bring the numbers of infection-fighting cells to a dangerously low level, leading to a life-threatening immune deficiency. Table 2 illustrates this connection between genes and proteins for several other diseases identified as candidates for gene therapy treatments.

TABLE 2. Gene Therapy Disease Candidates

Disease	Characteristics	Gene transferred	Action of gene product
Alzheimer's disease	Dementia occurring among older persons due to a loss of nerve cells in the brain. This process is associated with a buildup of protein plaques outside nerve cells and the accumulation of tangled protein fibers within nerve cells.	*Nerve growth factor, beta polypeptide (NGFB)*	Stimulates nerve cell function and prevents cell death.
Angina	Chest pain that occurs when the heart muscle is not receiving sufficient oxygen, due to buildup of cholesterol-based plaque in the coronary arteries that bring blood to the heart tissue.	*Vascular endothelial growth factor 2 (VEGF-2)*	Stimulates the growth of collateral blood vessels to provide the heart with additional oxygen.
Lung cancer	Development and growth of tumors in the lungs. Mutations in the tumor suppressor gene p53 are common in many cancers, including lung cancer.	*Tumor suppressor protein (p53)*	Provides functional p53 protein to correct mutation found in cancer cells. The p53 acts as a tumor suppressor by inhibiting division of cancer cells and stimulating programmed cell death, apoptosis.

TABLE 2. (continued)

Disease	Characteristics	Gene transferred	Action of gene product
Cystic fibrosis	A genetic disease characterized by breathing problems, respiratory infections, and difficulties with digestion due to the production of thickened secretions in the lungs and digestive system.	*Cystic Fibrosis Transmembrane Conductance Regulator (CFTR)*	Provides the functional protein to correct the genetic defect. CFTR controls balance of chloride ions between the exterior and interior of cells.
Diabetes	Excess of glucose in the blood due either to a deficit of the hormone insulin or the inability of body cells to respond to insulin. Damage to blood vessels leads to poor circulation, heart disease, stroke, kidney failure, blindness and nerve damage.	*Vascular endothelial growth Factor A (VEGFA)*	Stimulates the growth of new blood vessels to increase circulation in the legs.
Hemophilia B	An inherited bleeding disorder due to a deficit of factor IX, a protein that acts in the blood-clotting pathway.	*Factor IX*	Provides functional protein to correct the genetic defect.
Leukemia	A type of cancer characterized by uncontrolled growth of white blood cells. May result in anemia, bleeding problems, and enlargement of the liver and spleen.	*Interleukin 2 (IL2)*	Stimulates the response of the immune system against cancer cells.
Melanoma	A type of cancer developing from pigment-producing skin cells called melanocytes.	*Interleukin 24*, also known as *melanoma differentiation-associated gene (IL24, or mda-7)*	Acts as a tumor suppressor by inhibiting division of cancer cells and stimulating programmed cell death, apoptosis.

TABLE 2. (continued)

Disease	Characteristics	Gene transferred	Action of gene product
Duchenne muscular dystrophy	Hereditary, fatal illness characterized by muscle wasting and weakness due to a deficit in the muscle protein dystrophin.	Antisense Oligonucleotides that bind to *Dystrophin (DMD) RNA*	Block defective site in dystrophin RNA so that functional protein can be formed.
Prostate cancer	Development of tumors in the prostate gland, the second most common type of cancer in men. Besides the possibility of metastasis (spreading to other organs), prostate cancer can interfere with urination and sexual function. Prostate cancer cells have high concentrations of telomerase, an enzyme that maintains structures at the ends of chromosomes.	Vaccine for the enzyme telomerase	Stimulates immune system cells to recognize and destroy cancer cells that are producing telomerase in large amounts.
Severe Combined Immune Deficiency (SCID)	An inherited disorder leading to depletion of immune system cells that makes the person vulnerable to life-threatening infections.	*Adenosine Deaminase (ADA)*	Provides functional protein to correct the genetic defect.
Parkinson's disease	Problems with coordination and movement due to a deficit of the inhibitory neurotransmitter dopamine.	*Glutamic acid decarboxylase (GAD)*	Promotes the synthesis of gamma-aminobutyric acid (GABA), an inhibitory neurotransmitter.

ISOLATING A FUNCTIONAL COPY OF THE GENE

Out of thousands of human genes, how do we focus on the one that is causing a specific genetic disease? The sequencing of all human DNA, through the recently completed Human Genome Project, has yielded a wealth of information on the arrangement and structure of human genes. As a result of these studies, many human genes have been identified and **cloned**, isolated in a purified form. Cloned genes are complexed with DNA carriers called **vectors** in constructs called **recombinant DNA** molecules. Vectors contain DNA sequences that are compatible with a **host cell** into which the cloned DNA is introduced and duplicated. A marker gene makes an easily identifiable product and is often included to show that the vector has entered the cell. For the Human Genome Project, vectors were used to introduce human genes into bacterial cell hosts, since these cells are simple to grow in large numbers and can therefore produce significant amounts of human DNA for analysis. For gene therapy, it is necessary to transfer the gene to a vector that can deliver it to human cells.

How is a vector combined with the DNA it will carry? Molecular biologists make use of **restriction enzymes**, proteins that act like molecular scissors, which bind to double-stranded DNA and cut the strands at specific sequences. Many of these enzymes make staggered cuts that leave single-stranded regions at the ends. Using the same restriction enzyme on both the human DNA and the vector sequence will result in complementary or matching ends (Figure 3). Once the single-stranded ends associate with each other, the enzyme **DNA ligase** can be used to seal the gaps. Oftentimes specific DNA sequences are added to the vector in the regions surrounding the gene to control the production of the relevant protein.

While it is beyond the scope of this booklet to describe the methods used to clone human genes, it is worthwhile to understand how the cloned copy of a gene can be identified among a collection of recombinant DNA molecules, called a **gene library**. This library would contain thousands of clones, each one carrying a different DNA sequence. We can compare this process to looking for a needle in a haystack. Fortunately, molecular biologists can use a sort of "magnet," called a probe, to sift through the haystack of distracting genes. A probe consists of a short DNA sequence that matches a portion of the gene of interest. To understand how a probe helps to find a gene, it is useful to recall that DNA is a double-stranded molecule, as shown in Figure 4. Two chains of subunits called nucleotides associate with each other and twist to form a helix. Hydrogen bonds form bridges between the two chains, joining specific nitrogen-containing bases. An adenine (A) on one chain will bond to thymine (T) on the other, and a guanine (G) on one chain will bond to cytosine (C) on the other. A probe is designed to provide matching bases for part of the gene of interest. For example, if the gene sequence contains

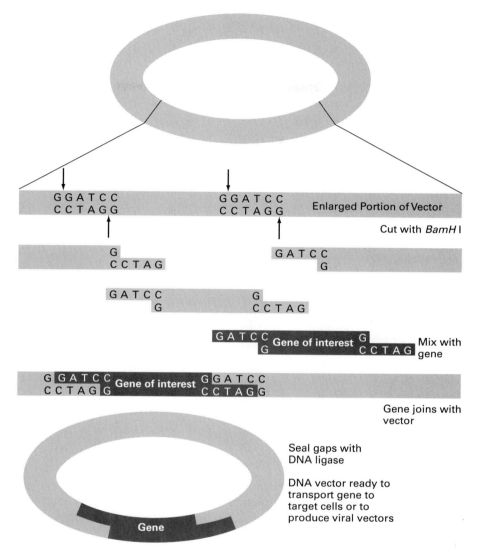

FIGURE 3. Adding a gene to a vector using the restriction enzyme *BamH* I.

AGCTGGAC then the probe could read TCGACCTG. Heat can be applied to cause the DNA strands to separate and when the mixture is cooled, the probe can attach to the gene sequence. Probes are labeled at one end so that their binding to the gene can be detected. The functional *ADA* gene can be located by using a gene library from a person who does not have SCID and a probe specific to a portion of the ADA gene sequence, as shown in Figure 5.

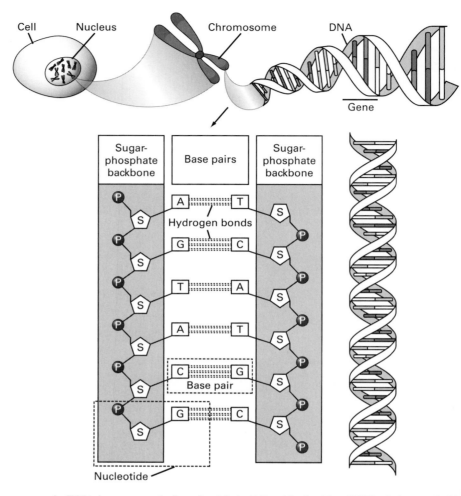

FIGURE 4. DNA is composed of nucleotide building blocks. Two DNA chains are held together by hydrogen bonds between specific pairs of nitrogen-containing bases.
Source: Palladino, *Understanding the Human Genome Project* 2e, Pearson Benjamin Cummings, Figure 2

INCORPORATING THE GENE INTO A CARRIER FOR GENE DELIVERY

Thanks to the Human Genome Project and many other gene identification studies, scientists have identified nearly all human genes, including thousands of disease genes. Finding a gene is an accomplishment, but a gene can only be useful for gene therapy if it can be placed into a cell and used to produce protein. The delivery method must be specific in targeting cells that are affected by the disease, efficient in allowing a significant number of cells to

```
G C T A T C G C G G G C T G C C G G G A G G C T A T C A A A    Portion of
C G A T A G C G C C C G A C G G C C C T C C G A G A G T T T    target gene


G C T A T C G C G G G C T G C C G G G A G G C T A T C A A A
                                                               Separate strands of
C G A T A G C G C C C G A C G G C C C T C C G A G A G T T T    target gene
```

```
              C C G A C G G C C C T  ←————————— Add labeled probe


G C T A T C G C G G G C T G C C G G G A G G C T A T C A A A    Probe binds to gene
C G A T A G C G C C C G A C G G C C C T C C G A G A G T T T    sequence
              C C G A C G G C C C T


A A T G C C C T T C G G C G A T G T T T T T T C T G G A G A
T T A C G G G A A G C C G C T A C A A A A A G A C C T C T      Probe will not bind
                                                               to other gene
A A T T G C A C T T G G A A C A G C A G C T C T G A G C C C    sequences
T T A A C G T G A A C C T T G T C G A C G A G A C T C G G G
```

FIGURE 5. Locating a copy of a specific gene using a probe.

receive the gene, and effective in transferring the gene to the cell's nucleus so that its sequence can be transcribed. Attempts to transfer "naked DNA," usually a recombinant DNA molecule with a cloned copy of the gene, have been inefficient. Both the cell's outer membrane and the membrane surrounding the nucleus act as barriers preventing DNA movement, while the cell's cytoplasm contains enzymes that degrade DNA. To improve gene delivery to human cells, disabled viruses are often used as vectors. This capitalizes on the natural ability of viruses to bind to cells and deliver the genetic material that they contain. To prepare a viral vector, genes that promote infection are removed and replaced with the gene of interest. In some cases, the gene delivered by the virus will be incorporated into one of the cell's chromosomes, but in others the virus remains as a separate entity in the nucleus. In either situation, the cell can use the newly acquired gene to produce the therapeutic protein.

For Andrew's gene therapy, a disabled mouse leukemia virus was used as a vector to introduce the functional *ADA* gene into a specific set of his own cells. This type of vector, a **retrovirus**, carries its genetic information in the form of RNA. It also has a unique enzyme called **reverse transcriptase** that converts the RNA strand to a DNA strand within the cell. The cell's enzymes produce the complementary DNA strand, and the resulting double helix can

integrate or be inserted into one of the cell's chromosomes as shown in Figure 6. Consistent with these properties of a retrovirus, producing viral particles to deliver the *ADA* gene requires a series of nucleic acid conversions, outlined in Figure 7 and illustrated in Figure 8. First, a DNA vector is constructed that contains the *ADA* gene along with a "packaging signal" that will allow the *ADA* sequence to be included in a viral particle after transcription. Then, these vector sequences are transcribed into RNA using a host cell "packaging line." This packaging cell also produces viral proteins, including reverse transcriptase and those of the viral core and surrounding envelope. Genes for these viral proteins have been integrated into the host cell chromosomes, in a manner that excludes their packaging into viral particles. The

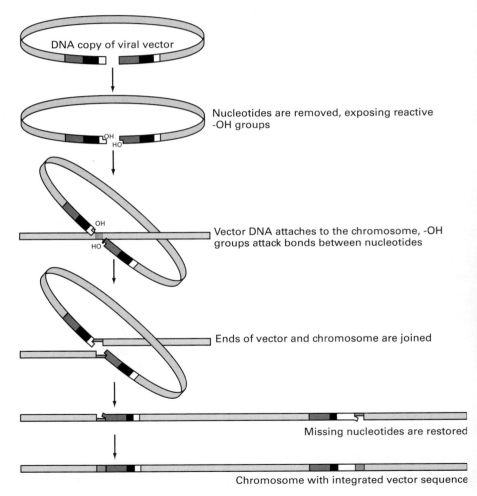

FIGURE 6. Integration of vector DNA into the chromosome.

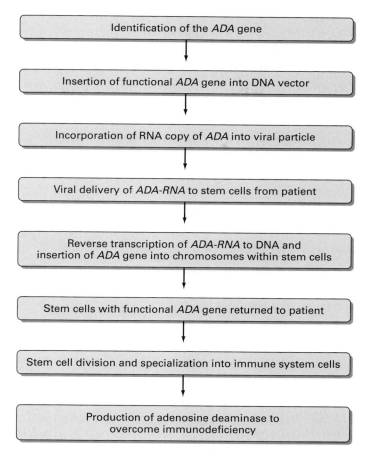

FIGURE 7. Preparing and delivering the ADA gene to treat ADA-SCID.

RNA copies of the *ADA* gene are then incorporated into viral particles assembled from the packaging line-derived viral proteins. For the gene therapy treatment, the viral vector attaches to a target cell and delivers its RNA copies of *ADA*. This information is converted to DNA by reverse transcription prior to integration into the chromosomal DNA.

For Andrew's treatment, the gene therapy target was a type of stem cell, an unspecialized cell capable of developing into one that could fulfill specific functions. In this case, the stem cell belonged to a blood cell lineage that would give rise to immune system cells throughout Andrew's lifetime. Dr. Kohn's team was able to collect these cells from the blood remaining in Andrew's umbilical cord immediately after his birth. The blood sample was enriched for possible stem cells by selecting those cells that carried a cell surface protein named CD34. The CD34+ cells were incubated in the presence of

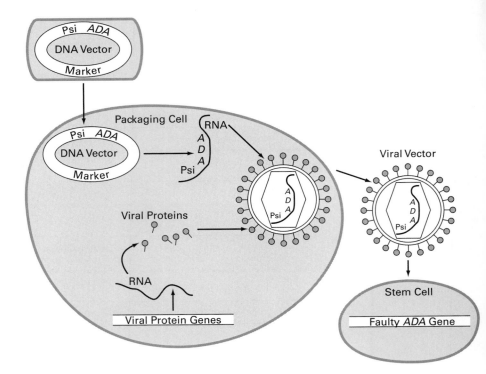

FIGURE 8. Production of a retroviral vector for delivery of the *ADA* gene. A DNA vector carrying the *ADA* gene is produced and transferred to a packaging cell. Transcription within the packaging cell yields RNA copies of the *ADA* gene along with a signal (Psi) for packaging into the viral particle. Viral proteins are produced from genes integrated into the chromosomes of the packaging cell. The viral vector delivers the RNA copy of the functional *ADA* gene to a stem cell.

the vector containing the *ADA* gene. They were also treated with a growth factor, so that their number would increase. *ADA*-RNA delivered by the viral particle would be converted to DNA and integrated into a chromosome within each stem cell. After a few days, these *ADA*-carrying cells were returned to Andrew's blood stream by injection (Figure 9). For the therapy to succeed, the stem cells would need to collect in the bone marrow and continue dividing, giving rise to immune system cells with the functional *ADA* gene. The researchers reasoned that these cells would have an advantage in repopulating Andrew's immune system because the restored ADA enzyme activity would promote DNA duplication and repair and relieve the toxic buildup of metabolites within the cells.

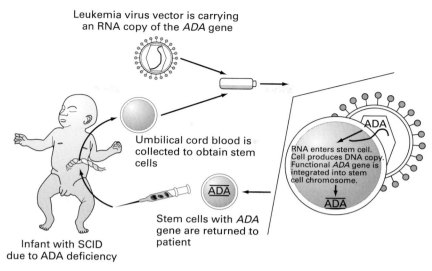

FIGURE 9. Introducing a functional gene for ADA with a mouse leukemia virus vector.

Which cells of the immune system need to be replenished for Andrew's recovery? Patients with ADA deficiency are lacking in three categories of immune system cells called lymphocytes: B cells, T cells, and natural killer cells. B cells mature to produce antibodies that mark viral and bacterial invaders for destruction. Helper T cells are important for communicating between cells of the immune system. Natural killer cells help eliminate infected cells, as do cytotoxic T cells. All of these cells are essential for fighting infectious agents that humans encounter on a daily basis.

DETERMINING WHETHER THE GENE PRODUCT IS MADE

How will physicians know that the newly introduced gene is functioning normally in the patient's body? Tests can be performed at many levels. DNA sequences associated with the vector can be detected so that the number of cells carrying the functional gene can be monitored. There are also ways to determine the level of the protein produced with biochemical tests. By both of these measures, Andrew's therapy looked promising. By the time Andrew was two years old, Dr. Kohn was able to show that cells containing the functional *ADA* gene had remained in his system. During this period, Dr. Kohn was also able to reduce the level of ADA provided through protein replacement therapy. Andrew's body responded by producing more of the *ADA*-bearing immune system cells. Although the gene had populated only 0.01% of Andrew's stem cells, between 5 and 7% of his circulating white blood cells were shown to have the functional *ADA* gene. Yet the most crucial indication of the gene's action is whether the patient begins to recover from the illness. By continuing to reduce the ADA dose supplied by injections, Dr. Kohn would be able to determine whether the number of cells carrying the functional gene would be sufficient to maintain Andrew's health.

Partial Successes and Ongoing Trials

GENE THERAPY FOR SEVERE COMBINED IMMUNE DEFICIENCY (ADA-SCID)

Prior to Andrew Gobea's treatment as an infant, a gene therapy trial aimed at helping Ashanthi DeSilva, a four-year-old girl with ADA-SCID, was successful at partially restoring her immunity (Figure 10). In this treatment, viral particles were used to introduce *ADA*-RNA directly into white blood cells of the immune system. Since these cells would have a limited lifetime and the patient would continue to replace them with cells carrying the defective gene,

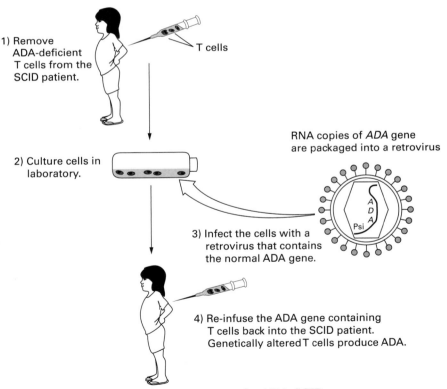

FIGURE 10. The first gene therapy treatment for ADA-SCID.
Source: Thieman and Palladino, *Introduction to Biotechnology*, Pearson Benjamin Cummings, Figure 11.12

she needed to undergo repeated gene therapy treatments to maintain an appropriate level of ADA in her body. A low dose of ADA enzyme replacement therapy was also continued in this case. As a result of this combination of treatments, the patient is living an active and healthy life (Figure 11). This early success encouraged scientists to find ways of improving gene therapy. The goal of achieving a more permanent effect led Dr. Kohn and other researchers to try introducing *ADA* genes into stem cells. This effort would have the added advantage of generating a set of cells capable of defending against a wider range of invaders than the subset of white blood cells treated with each repetition of the transfer procedure.

GENE THERAPY FOR CYSTIC FIBROSIS

Initial successes have also been demonstrated in gene therapy treatments for other genetic conditions. Since the disease **cystic fibrosis** affects the epithelial cells lining air passageways to the lungs, access to these cells provides an advantage for gene delivery. One of the manifestations of cystic fibrosis is a buildup of sticky mucus in the airways as a result of a defective ion channel protein, called cystic fibrosis transmembrane conductance regulator (CFTR). The protein normally regulates the balance of chloride ions on either side of

FIGURE 11. Gene therapy patient Ashanthi DeSilva at age 6, two years after treatment. *Source:* Time Life Pictures/Getty Images

epithelial cell membranes. Patients with cystic fibrosis make an altered version of this protein that is delayed in its transport to the cell membrane. The resulting mucus accumulation on the cell surfaces leads to infections and lung damage. For the gene therapy, drops containing a modified adenovirus vector were used to deliver the ion channel protein gene to a small number of airway cells in a patient's nose or lungs. The gene became active in these cells, producing the RNA that is used to make the ion channel protein (Figure 12). This beneficial effect lasted for 30 days. Efforts are now focused on methods to deliver the gene to a greater number of cells and to prolong the duration of gene activity.

GENE THERAPY FOR CANAVAN DISEASE

Gene therapy for **Canavan disease** has also shown promising results. This condition occurs when the gene for an enzyme called **aspartoacylase** (ASPA)

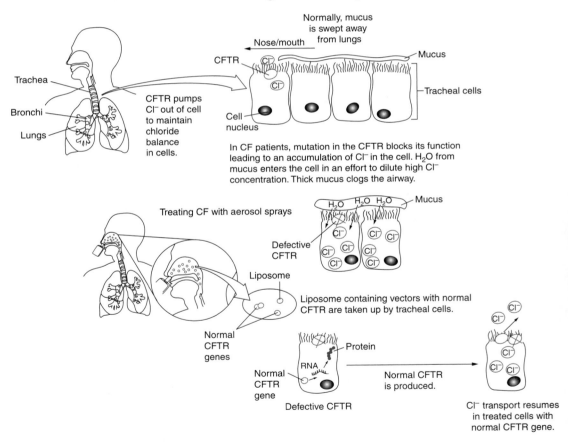

FIGURE 12. Gene therapy for cystic fibrosis.
Source: Thieman and Palladino, *Introducton to Biotechnology,* Pearson Benjamin Cummings, Figure 11.13

is faulty. ASPA is usually produced by **oligodendrocytes**, cells in the brain that synthesize **myelin**. Myelin forms an insulating covering on nerve cells called **neurons** to insure rapid transmission of nerve signals. ASPA breaks down a chemical compound called N-acetyl-aspartate that is produced by neurons. One of the resulting products, acetate, is required for myelin synthesis. In Canavan disease, the lack of ASPA causes a buildup of N-acetyl-aspartate, interfering with myelin production. Children inheriting this disease show early developmental delays that affect their vision, speech, and movement. They are unable to crawl or support their heads. The impairment of nerve transmission leads to death between the ages of three and ten. Since the enzyme defect is due to a change in the DNA sequence of a single gene and there are no other successful treatments, patients with Canavan disease seemed to be ideal candidates for gene therapy. The delivery of the ASPA gene to the brain was attempted in two ways. With one method, genes on a nonviral vector were packaged with **liposomes**, spheres composed of lipids, and introduced surgically into the skull (Figure 13). The lipid component was designed to merge with cell membranes to introduce the genes into nervous system cells. There were indications of ASPA activity and some short-term improvements in these patients. In a second group of studies, a viral vector specific for neural cells was used to deliver the genes. Improvements were again observed, especially for the youngest patients. For a two-year-old girl treated by this method, myelin levels in brain neurons have been restored. She has gained control of arm and leg movements and shown improvements in vision and speech.

Do all of these promising signs mean that gene therapy is ready to move out of the experimental stages and into regular clinical practice? Unfortunately, a series of setbacks has clouded the path and necessitated slower and more careful steps.

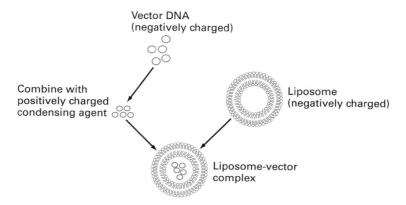

FIGURE 13. Association of vector DNA with a synthetic sphere of lipids called a liposome.

Setbacks

IMMUNE REACTIONS TARGETING CELLS CARRYING THE VECTOR

In 1999, a test to determine the safety and dosage of genes for treating **Ornithine Transcarbamylase (OTC) deficiency** raised questions concerning responses to the vectors selected to transfer therapeutic genes. OTC is an enzyme that participates in a pathway converting ammonia, a by-product of using the amino acid subunits of proteins for energy, into urea for elimination in the urine. Unless ammonia-eliminating treatment is started at birth, newborns with OTC deficiency fall into an irreversible coma within days as the fast-paced buildup of ammonia affects their brain cells. Gene therapy seemed to be the only hope for a disease with such rapid effects. At the University of Pennsylvania, a modified adenovirus vector was selected to deliver the *OTC* gene to the livers of adults with the disease and to carriers of the disease.

One of the volunteers was an 18-year-old named Jesse Gelsinger, a patient with a mild form of OTC deficiency whose symptoms had been minimized by a low-protein diet and ammonia-binding drugs (Figure 14). At first, Jesse's response with an elevated body temperature was similar to that of other patients in the trial, but the situation soon became life threatening. In addition to entering the targeted liver cells, the vector also entered immune system cells, and signals released from these cells triggered a massive immune response. This response was primed by a previous infection with a similar virus. Jesse experienced a severe blood clotting reaction, leading to both kidney and lung failure. His death drew attention to the need to check for other medical conditions that could affect the response to vectors used in gene therapy. New regulations concerning recruitment of volunteers for gene therapy trials and for reporting adverse effects were put into place.

INADVERTENT ACTIVATION OF CANCER-CAUSING GENES BY A VECTOR

Even in patients without prior exposure to vector-like viruses, the effects of the vector can be harmful. This was observed when gene therapy treatment was applied to another form of SCID, called **X-SCID**. The X-SCID immune deficiency is the result of a defect in the *Interleukin 2 Receptor Gamma Chain (IL2RG)* gene carried on the X chromosome. The product of the *IL2RG* gene is integral to the maturation of immune system cells in response to chemical signals released during infections. Patients with X-SCID have defective B cells and are lacking in T cells and natural killer cells. During a clinical trial conducted in 2000 in France, a retroviral vector was used to deliver a functional *IL2RG* gene to bone marrow stem cells (CD34+ cells as

FIGURE 14. Gene therapy volunteer Jesse Gelsinger died as the result of a reaction to the vector delivering therapeutic genes.
Source: AP Photo

in Andrew's therapy) of patients with X-SCID. Initial results of this treatment showed substantial promise. The gene therapy reconstituted the immune systems in nine of the ten patients, with a wide variety of immune system cells carrying the functional *IL2RG* product. This was the first instance of gene therapy correcting a disease without the need for supportive medications or repeated treatments. Several years later, however, the treatment showed damaging side effects when three patients developed T-cell leukemia. The leukemia was traced to the insertion of the vector with the therapeutic gene near the start site of a gene called *LMO2*. The *LMO2* gene product naturally stimulates the growth and maturation of immune system cells. Sequences on the viral vector enhanced the production of LMO2 and led to uncontrolled growth of these T cells. This type of genetic change as a result of vector integration is described as **insertional mutagenesis**. The outcome became even more tragic when one of the patients died as a result of the T-cell leukemia.

While these circumstances may have been unique to treatment of X-SCID, the possibility of insertional mutagenesis was considered to be sufficiently dangerous for a moratorium to be placed on gene therapy with retroviral vectors until the safety of related gene therapy trials could be evaluated. Decisions on the type of vector and where it will be directed to integrate into the chromosome are crucial for the safety of gene therapy treatment.

INSUFFICIENT NUMBERS OF CELLS PRODUCING THE GENE PRODUCT

The vector used to deliver *ADA* genes to Andrew's stem cells was a mouse leukemia virus intended for integration into his chromosomes. Andrew has not developed complications related to the vector, but will the *ADA*-bearing cells restore his health? At age four, Andrew seemed well enough to try eliminating his ADA injections to test whether he could make sufficient amounts of enzyme on his own. Within two months of reducing his ADA dose, however, Andrew began losing weight and suffering from two infections, indicating that his immune system failed to eliminate disease-causing invaders. For Andrew's safety, the ADA injections were resumed. The number of genes delivered was not enough to reprogram Andrew's immune system to make health-promoting levels of ADA on its own. The researchers concluded that delivering the gene to a greater number of stem cells, which would then give rise to an even larger number of circulating white blood cells, would be essential to successful gene therapy for this disease. Andrew has been maintained on ADA injections since 1993 and is approaching his teen years in relatively good health. His immune system, however, has less than 10% of the typical number of T cells needed for optimal function and there is concern that this might have an adverse affect later in his life.

Challenges

Clearly, significant challenges remain on the path toward successful gene therapy. To overcome the obstacles so far encountered, scientists must find ways to improve the therapy in three main areas. Safe and effective delivery of the gene is of initial importance. In addition, the procedures must ensure that a sufficient amount of the gene product is produced to overcome the effects of the disease. To achieve a lasting improvement for the patient, methods must be found to increase the persistence of the gene so that it can continue to deliver its health-promoting product. Fortunately, advances are being made in all of these areas.

SAFE AND EFFECTIVE DELIVERY OF GENES
Improving the Action and Safety of Vectors

The effectiveness of gene therapy hinges on the ability to deliver the gene to a specific set of cells that require its activity. While vectors are still a method of choice for this task, scientists are constructing improved versions to increase efficiency and remedy some of the problems that surfaced in early gene therapy trials.

In order to understand the options available for gene delivery, it is helpful to review the variety of vectors currently available to researchers (Table 3). The simplest type of vector is a **plasmid**, a small circle of DNA found outside the chromosome of bacterial cells. These vectors have been useful for isolating and duplicating human genes but do not persist in human cells. The earliest gene therapy vectors were derived from retroviruses that carry RNA as their genetic material. These include the mouse leukemia viral vector used to transfer the *ADA* gene to Andrew's umbilical cord stem cells. Retroviral vectors were favored for promoting chromosome integration of the therapeutic gene but were limited to producing permanent change only in growing and dividing cells. **Lentiviral vectors**, also from a retroviral source, were then considered for the possibility of delivering materials to a wider range of cell types. When chromosomal integration was not required, **adenovirus vectors**, composed of DNA, were employed for diseases that would benefit from genes with transient effects. Adenovirus causes respiratory illness in humans and is recognized by the immune system, as tragically shown in Jesse Gelsinger's case. Recent attention has focused on DNA-containing **adeno-associated viruses** (AAV), which were discovered during adenovirus infections. A large number of humans have been naturally exposed to AAV without harmful effects. As it requires a helper virus in order to duplicate within cells, the infectivity of AAV-based vectors should be minimal.

One trend in vector construction is to eliminate as many of the viral sequences as possible. This has three significant advantages. The vector can carry a larger amount of DNA to help correct the genetic defect. Unexpected effects of vector sequences inside human cells would be reduced. The chances of reconstituting the virus in its infectious state within the body would also be minimized. This has led to second and third generation versions of adenovirus and lentivirus vectors that have highly reduced viral sequences. Eliminating most of the viral sequence has led to "gutted" versions of adenovirus vectors with the goal of minimizing possible immune

TABLE 3. Guide to Gene Therapy Vectors

Vector	Composition	Chromosomal integration	Length of expression	Size of insert	Advantages	Disadvantages	Gene therapy use
Adeno-associated virus	Single-stranded DNA	Integration site on chromosome 19	Transient	3.5–4.7 kb*	Non-immunogenic, enters brain and muscle cells	Small amount of DNA transferred	Canavan disease, Hemophilia B, Parkinson's disease
Adenovirus	Double-stranded DNA	Normally does not integrate; helper-dependent viruses have shown random integration	Transient	Up to 30 kb	Carries large genes	Immune reactions either upon treatment or readministration	Cystic Fibrosis, OTC deficiency, glioma (in cell lines), p53 therapy for cancer
Lentivirus (e.g., HIV)	Single-stranded RNA	Random integration within active genes	Long term	8 kb	Delivers genes to nuclei of nondividing cells	Insertional mutagenesis	Antisense therapy for HIV
Plasmid	Double-stranded DNA	Normally does not integrate	Transient	5–10 kb	Useful for isolating disease-causing genes	Unable to deliver genes to human cells	Carries therapeutic gene into packaging cell for retroviral and lentiviral vector applications
Retrovirus (e.g., mouse leukemia virus)	Single-stranded RNA	Random integration near start sites of active genes	Long term	7.5 kb	Delivers genes to nuclei of growing and dividing cells	Insertional mutagenesis	ADA-SCID, X-SCID, chronic granulomatous disease

*kb = kilobases = 1,000 nucleotides of RNA or DNA for a single stranded vector or 1,000 base pairs of DNA for a double-stranded vector responses. Modifying a specific sequence at one end of the lentiviral RNA has the unique effect of producing a **self-inactivating** (SIN) vector. During reverse transcription, this region is duplicated onto both ends of the DNA copy of the virus, preventing transcription of the viral vector sequence after it is integrated into the host chromosome.

The efficiency of delivering genes by different viral vectors is dependent upon the type of cell infected. Genes carried by retroviral vectors can only successfully reach the chromosomes of a growing and dividing cell since the temporary disruption of the membrane surrounding the cell's nucleus is required for the viral DNA to have access to the cell's genetic material. To infect nondividing cells, scientists are proposing to use lentiviral vectors that can traverse the nuclear membrane. This would provide an advantage for delivering genes to nerve cells, for example. These vectors can integrate DNA copies of their RNA genetic material into the chromosomes of either a dividing or nondividing cell. It is also possible to produce hybrid viral vectors that have the characteristics of one virus on the surface for compatibility with a specific type of body cell and the genetic characteristics of another virus for insertion into the cell's chromosomes.

Since the Human Immunodeficiency Virus (HIV) is one virus in the lentiviral category, there are concerns about additional health risks if the viral vector should be able to reconstitute its AIDS-producing characteristics during or following gene therapy treatment. There are many methods that attempt to prevent disabled viruses from recapturing their infectious abilities. Frequently these vectors are constructed in replication-deficient states. They are missing genes that would be needed to produce fully functional viral particles within the cell. The gene to be delivered is included with a packaging signal and **long terminal repeats** (LTR), sequences at both ends of the vector containing start and stop signals for gene expression. As shown in Figure 15, the vector is produced in the laboratory within a packaging cell host that also contains viral protein genes provided by one or more helper viruses or with copies integrated into the host cell's chromosomes. These genes lack a packaging signal. In this way, only the therapeutic gene and sequences needed for its chromosomal integration and transcription are packaged inside the viral particles. Once the virus enters the patient's cell and releases this genetic material, it does not have the capability of producing new viruses. While the viral vector could be converted to a natural and infective version of the virus by combining with a natural virus within the host's body, the vector is constructed so that the number of combination events required would be large enough to essentially eliminate this possibility.

Innovations in the numbers and arrangements of genes that can be included within the viral vector sequences have been advanced. More than one gene can be included in some vectors capable of carrying a large genetic

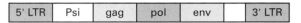

A. Structure of the lentivirus genome. LTR = long terminal repeat containing promoter sequence for transcription initiation. Psi = packaging sequence needed for insertion into viral particle. Gag = region with genes for viral proteins forming core particle that surrounds genetic material. Pol = region with genes for the viral proteins reverse transcriptase, endonuclease, protease, and integrase. Env = region with genes for outer viral envelope.

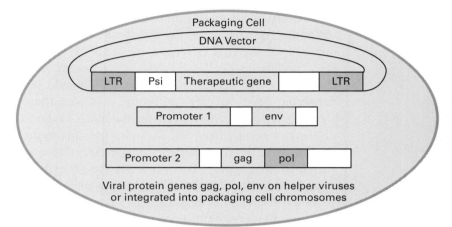

B. Genes for producing the viral vector are separated into three groups to avoid production of infectious viruses. Only the therapeutic gene is associated with the packaging signal, Psi. Infectious virus is not produced since multiple rearrangements would be required to restore the viral genome. Promoter = start signal for transcription.

FIGURE 15. Methods used to increase the safety of lentiviral vectors.

payload. The genes are arranged in opposite orientations, each with its own start signal, called a **promoter**. Some unique constructs even allow many genes to be utilized from the start signal.

Controlling the Timing of Gene Expression

Controlling when and how efficiently a new gene acts within a cell has advantages for avoiding side effects as well as increasing the level of therapeutic protein produced. Genes on retroviral vectors have historically been under the control of LTR sequences at the ends of the vector. The promoters in these regions may not be as efficient as those that precede the gene in its normal chromosomal setting or may cause an overproduction of the protein due to enhanced efficiency. To make synthesis of the gene product more specific in the treatment of ADA-SCID, the promoter region that is normally associated with the *ADA* gene could be included in the vector that is delivering the *ADA* gene to stem cells. Some promoters act as start signals only under certain environmental conditions. If these were included on the gene delivery vector, then the gene would be silent until those conditions could be applied. This would allow physicians to switch the medicine-producing capabilities of the cell on and off as needed. Another alternative is to include a "suicide gene" that would make the cell sensitive to a specific drug. If the cells with the new gene show abnormal behavior, the suicide gene could be switched on, and administering the drug could eliminate the cells.

Controlling Vector Integration

Choice of the vector can greatly influence where possible integration events occur. Recent studies have shown that both Mouse Leukemia virus and HIV virus integrate in active gene regions while the Mouse Leukemia virus appears to preferentially integrate near the start sites of these genes.

The ability to control the site of insertion of the vector into the chromosome would present a significant safeguard for the patient's health. One approach is to use methods that target specific sites on chromosomes for gene insertion. For example, the adeno-associated virus (AAV) is known to integrate at a specific site on chromosome 19. An AAV vector has been used to successfully deliver the gene for a blood-clotting factor, **Factor IX**, so that patients showed correction for the blood-clotting disease **Hemophilia B**. An additional example involves the use of a viral integrase system. Integrase is an enzyme that controls where a virus will insert in a host chromosome. Sequences within human cells have been found that resemble those targeted by a viral integrase. If the gene for this integrase is included on a plasmid vector, the vector sequence can be inserted in specific positions recognized by the integrase. This method has been used to successfully transfer genes for Factor IX into mouse chromosomes with the resulting production of significant amounts of the blood-clotting factor. Since many gene transfers involve removal of the target cells from a person's body and return after treatment with the vector, it is possible to determine whether the insertions are in safe locations prior to replacing the cells.

Avoiding Adverse Effects on Chromosomal Genes

Minimizing the action of the vector on genes near the insertion site is also a possibility. The LTRs of retroviral vectors act as **enhancers,** increasing transcription of chromosomal genes on either side of the insertion site over long distances. Vectors can be constructed that will minimize these enhancers to avoid producing large amounts of products from genes in the vicinity of the insertion. It is also possible to include chromosome insulators in the vector construct. These sequences limit the action of enhancers and could shield chromosomal genes from enhancers found on the viral vector.

In order to avoid specific interaction between the vector and chromosomal sequences, it is possible to construct a vector that would remain separate from the chromosomes. In order to insure its stability within the cell and its transmission to the cell's descendants, this vector would need chromosomal elements such as the **telomeres** found at the ends of chromosomes and a **centromere** for attaching to fibers involved in chromosomal divisions. This type of artificial chromosome has already been constructed for use in yeast cells so the possibility of a "47th" human chromosome is not such a far-fetched scenario. The difficulty would be in delivering such a large amount of DNA to the cell.

Nonvector Gene Carriers

If the adverse effects of using a vector are a concern, other methods of transporting the gene can be implemented. For example, the therapeutic DNA can be introduced into liposomes for delivery into the cell when these membrane sacs fuse with the cell's membrane. We have seen that liposome delivery provided some improvement in one of the trials for treating Canavan disease. This method was also used to treat lung cancer by delivering the gene for p53, a tumor suppressor protein. The p53 protein normally slows cell division when damage to DNA has occurred and participates in a pathway that leads to cell death when this damage cannot be repaired. Since *p53* is a gene that represents the most common type of mutation in a wide variety of cancers, the potential to correct this defect in tumor cells can have a significant impact on an insidious disease. DNA can also be encased in microscopic nanoparticles that can interact with cells. Nanoparticles made of silica can deliver their genetic payload to cells before being detected by the body's immune system. These nanoparticles have amino groups that bind to DNA and the arrangement of these groups can be selected to optimize gene transfer to specific cell types. Nanoparticles have already been used to deliver copies of the *p53* gene in treating cancer in mice. A third alternative is to induce a short-term instability in the cell's membrane through the use of electric shock, a treatment called electroporation. This method has been used to deliver a gene for blood-clotting **Factor VIII** to fibroblast cells that were then injected into patients. These patients had short-term relief from **Hemophilia A** symptoms.

Choosing the Route for Gene Delivery

In concert with the type of DNA carrier selected for transferring the genes, the route of gene delivery must be considered. *Ex vivo* delivery is a method for adding genes to cells outside the body and then infusing the cells into the patient. This is the technique that was used to treat white blood cells and stem cells taken from and returned to SCID patients. *Ex vivo* delivery is also applicable to stem cells from donors. Multipotential stem cells that can give rise to more than one type of specialized cell can be particularly useful in the treatment of hemophilia. One such cell line gives rise to blood-forming cells as well as liver cells, providing two locations where clotting factors can be produced. Stem cells that can be induced to differentiate into neurons that produce the neurotransmitter dopamine can be used to correct the defect in Parkinson's disease. Adding genes that guide and enhance the process of stem cell specialization has been shown to optimize this process in animal systems. Stem cells that normally assist in the wound-healing process have been programmed to deliver a fatal blow to cancer cells. These mesenchymal stem cells carrying a gene for human interferon were shown to localize in tumors and destroy cancer cells in mice.

In addition to *ex vivo* delivery, two other major methods of gene transfer are available. *In situ* gene therapy targets a small group of cells or a small area of the body. This process has been effective for treating skin cancer by injecting genes directly into affected surface tissue. *In vivo* therapy, as used to deliver OTC genes to the liver, transfers genes into the body directly where widespread distribution is possible. Figure 16 shows a comparison between *ex vivo* and *in vivo* gene delivery. It has also been proposed that delivering therapeutic genes to fetuses in the uterus (*in utero* delivery, a specialized case of *in vivo* transfer) would provide significant advantages for treating specific

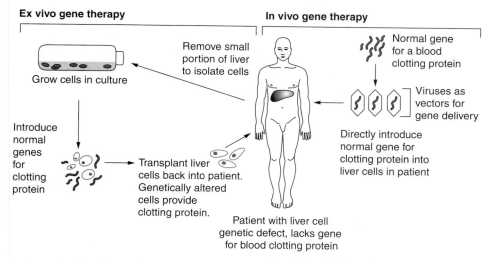

FIGURE 16. Comparison of *ex vivo* and *in vivo* methods of delivering therapeutic genes.

Source: Thieman and Palladino, *Introducton to Biotechnology,* Pearson Benjamin Cummings, Figure 11.10

diseases. Cystic fibrosis could be treated by adding genes to the amniotic fluid since the genes would come in contact with airway cells when the fetus takes in and expels the fluid. Access to the liver via fetal circulation could provide a way of treating hemophilia. Treating Duchenne muscular dystrophy *in utero* could allow correction of the defect in a wider array of muscle groups than those accessible after birth. Stem cell transplantation into a 14-week-old fetus with X-SCID was shown to restore normal levels of T cells and natural killer cells before birth. *In utero* therapy could also minimize immune responses to newly introduced proteins as the genetic material is introduced at a time when the fetus is becoming tolerant of body proteins.

PRODUCTION OF A SUFFICIENT AMOUNT OF GENE PRODUCT

The advantage of specific gene delivery could be compounded by a method that allows the number of corrected cells to increase within the body. For ADA-SCID gene therapy, Italian researchers are experimenting with a way to populate the bone marrow with a larger percentage of cells carrying the functional *ADA* gene. They treat the patient with a chemical that specifically reduces the number of stem cells in the bone marrow, allowing the newly delivered *ADA*-carrying cells to divide and replace them. Immune function was restored for two patients treated at the ages of 7 and 30 months. The children are living healthy lives at home with their families, without the need for ADA enzyme replacement therapy. Unlike the low numbers of T cells seen in Andrew's system, these patients have adequate numbers of T cells as well as antibody-producing B cells. Scientists in the United States, including Dr. Kohn, are undertaking clinical trials with this methodology.

A similar process for removing some of the patient's defective cells has also been effective in a gene therapy treatment for chronic granulomatous disease. In this rare disorder, patients lack one subunit of an enzyme complex called phagocyte NADPH oxidase (phox). They are subject to chronic infections because this enzyme is essential to the function of **phagocytes**, white blood cells that engulf and destroy bacterial and fungal invaders. In addition to delivering a functional gp91-*phox* gene to phagocytes called neutrophils, German clinicians used a low dose of chemotherapy to eliminate faulty white blood cells before returning the neutrophils to two patients. The treatment allowed the patients to recover from ongoing infections and stay free of new infections for a year and a half. This hopeful news is tempered by the discovery that the neutrophils were accumulating at increased levels within both patients, causing concern that leukemia may develop as a result.

To restore a patient's health, how much of the gene product must be derived from the functional genes? While the answer will vary according to the disease being treated, the amounts have been surprisingly small in some cases studied to date. For ADA-SCID, a significant benefit can occur by restoring the ADA levels to 10% of those in people without SCID. Hemophilia symptoms have been alleviated with as little as 2% of the normal clotting factor. This shows that while gene therapy protocols need to be optimized beyond current achievements, the increment needed for substantial benefits may be within reach.

ACHIEVING A LASTING IMPROVEMENT FOR THE PATIENT

Achieving and maintaining full expression of a newly introduced gene depends upon multiple characteristics of the chromosomal location into which it integrates. Some chromosomal regions show high levels of gene activity whereas others are relatively quiet. Enhancer sequences located at long distances on either side of the gene can cause an increase in transcription. Alternatively, regions called silencers can interfere with RNA production. Natural mechanisms for controlling gene activity will also be applicable. Methylation of cytosine residues near start sites is a way of preventing transcription or "silencing" specific genes. One concern of gene therapy researchers is that therapeutic genes would be silenced over time in the cell types that carry them. Methods for removing methyl groups are highly generalized so it is not presently possible to selectively restore a specific inactivated gene to its active state.

Treatment of many diseases requires that the genetic correction be permanent. Cells carrying the gene must persist, evading detection by the immune system and avoiding elimination from the cell population by senescence, cell aging. However, gene therapy can also be applied to medical conditions that benefit from short-term exposure to a functional gene product. Some diseases interfere with development at certain times in the life cycle and harmful effects could be eliminated if active genes were available. Other treatments are only needed when the effects of a disease are at their height. For example, gene therapy could be used to eliminate cancer cells and would not be needed thereafter.

Future Approaches

CORRECTION OF A PATIENT'S DNA

To this point, most gene therapy has focused on adding a functional copy of a gene and relying on its activity to compensate for the faulty one. Further refinements of the process would allow removing, replacing, or repairing the faulty gene for a more targeted cure. DNA binding proteins called **zinc-finger**

proteins (ZFP) can be engineered to recognize and bind to regions near a defective gene (Figures 17 and 18). In recent experiments using human cells in the laboratory, enzymes attached to these proteins have been shown to specifically cut the DNA in the region of the faulty *IL2RG* gene that causes X-SCID. The gene was restored to its functional state when scientists provided a small stretch of DNA that carried the correct DNA sequence. This method is analogous to sending a microscopic surgeon into the cell to repair the damaged gene. Besides gene repair, this method can also be used for gene regulation, either

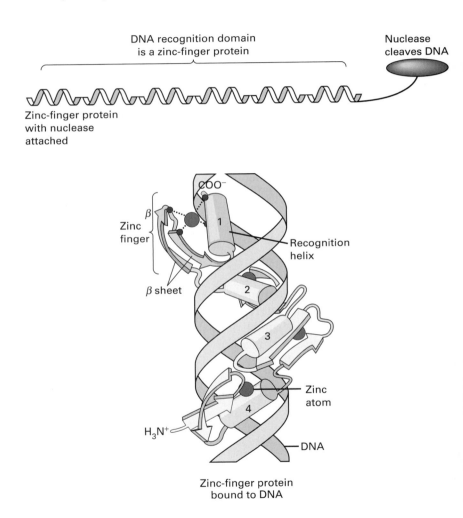

FIGURE 17. Structure of a zinc-finger nuclease used to repair genes. The enzyme is composed of a zinc-finger protein that recognizes and binds to a specific DNA sequence and a nuclease that will cleave DNA in that region.

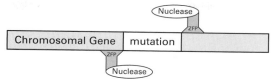

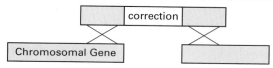

FIGURE 18. Use of a zinc-finger nuclease to repair genes. Zinc-finger proteins (ZFP) with attached nucleases bind to DNA regions surrounding a mutation and nucleases cleave the DNA to remove the defective area. A corrected DNA sequence replaces the damaged version, producing a functional gene.

switching on genes that would be helpful to a patient or switching off disease-causing genes. Efforts are underway to test the safety and efficacy of these methods in patients with peripheral artery disease and diabetic neuropathy.

INTERFERING WITH THE PRODUCTION OF THE GENE PRODUCT

Efforts to improve gene therapy can extend beyond events occurring at the DNA level. In some cases, it is possible to interfere with the action of a faulty gene by preventing its gene product from forming. Two of these methods reduce the activity of RNA, the intermediate in the production of proteins. RNA has chemical similarities to DNA, but it works as a single chain. In some cases, the nucleotides within the chain join to others on the same chain by hydrogen bonding to produce a folded molecule. As shown in Figure 2, in order to make the protein product of a specific gene, the DNA is first used as a pattern to produce messenger RNA (mRNA). Messenger RNA carries the instructions for the order in which to arrange the amino acids in the protein. The cell's protein-synthesizing organelles, the ribosomes, recognize this mRNA and provide a surface upon which the amino acid sequence of a protein can be built. The cell's amino acid carriers, transfer RNAs (tRNAs), deliver amino acids to the ribosome, binding briefly with areas on the mRNA that match the position where specific amino acids will be placed. The mRNA is often referred to as the **sense** sequence because of its capacity to direct the production of a specific protein. Scientists can direct the synthesis of **antisense RNA**, a molecule with a complementary sequence that can join to the sense chain by hydrogen bonding (Figure 19). This double-stranded sense–antisense combination cannot be recognized by the ribosome, preventing production of

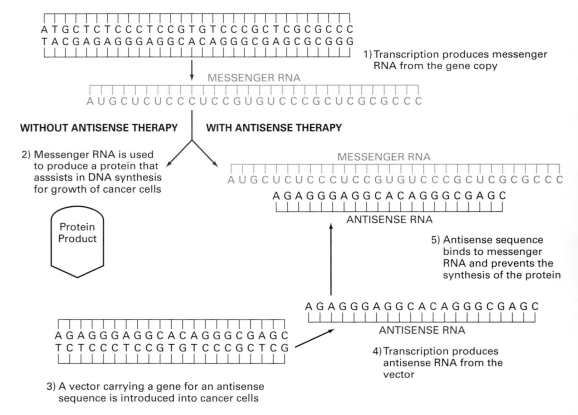

FIGURE 19. Use of antisense technology to prevent gene product formation in cancer cells. The bottom strand is the template for each RNA molecule produced.

the protein from the sense chain. In combination with chemotherapy, antisense technology has shown promise for reducing tumors in patients with kidney cancer (Figure 19). Initial success in lowering the amount of HIV in patients with AIDS has also been achieved with antisense therapy.

RNA interference represents another method of stopping protein production by selectively degrading mRNA molecules. An enzyme degrades double-stranded RNA to produce **small interfering RNA molecules** (siRNAs). These dissociate into single strands that bind to mRNA molecules and mark them for destruction by cellular enzymes (Figure 20). This strategy is proposed as a way to treat Huntington's disease if a specific difference between the disease-causing and normal forms of the gene can be targeted.

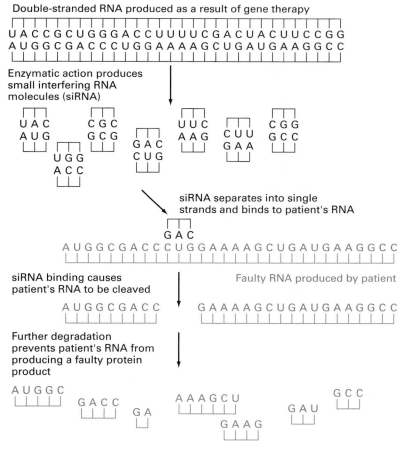

FIGURE 20. Use of RNA interference to prevent the product of a gene from forming within a cell.

Avoiding breakdown of these siRNAs during delivery has been difficult. One novel approach is to link them to cholesterol, a lipid that can assist the siRNAs with cell entry due to its natural association with the cell membrane. Scientists used these siRNAs to interfere with the production of apolipoprotein B (apoB), a protein needed to produce carriers that transport cholesterol in the bloodstream. When tested in mice, this method lowered cholesterol levels, showing successful interference by the RNAs. Use of a viral vector to deliver interfering RNAs was successful in reducing the expression of alpha-synuclein in rat brains; this could be used as a possible gene therapy for Parkinson's disease. Preventing a faulty gene from having its effect within a cell, either by antisense technology or RNA interference, has the potential for relieving the symptoms of a wide array of genetic diseases.

Several other processes offer the possibility of correcting the genetic defect at the RNA level. One of these, called **trans-splicing**, capitalizes on a natural process that removes extraneous sequences from RNA. Human genes are interrupted by non-protein-coding sequences called **introns**. These introns divide the gene region into segments called **exons**, the protein-coding regions of a gene. Both introns and exons are copied into RNA but a process called **splicing** removes the introns and joins the exons before the protein is produced (Figure 21). This process is normally carried out by a complex of proteins called a spliceosome that recognizes specific sequences at the junctions between introns and exons. Splicing usually occurs along the same RNA molecule, using a process called cis-splicing. But in some cases, splicing can occur between different RNA molecules, in which case it is referred to as trans-splicing. To produce a therapeutic effect, researchers would provide an RNA molecule with a corrected exon for the region that is faulty in the patient's gene. Trans-splicing would produce a fully functional RNA for protein synthesis by combining the patient's RNA with the newly introduced RNA (Figure 22). This method has been used to correct a Factor VIII deficiency in mice. It will be of significant advantage for correcting large genes, as only part of the RNA needs to be provided to give the therapeutic effect.

An alternative trans-splicing-correcting mechanism involves the use of a type of RNA that behaves like an enzyme. These **ribozymes** assist with splicing reactions by recognizing intron/exon junctions, removing introns and connecting exons together. In a novel approach, ribozymes were used to cut defective versions of a mouse RNA and then add a corrected region during the splicing step. In human cell lines, this approach has been successful in correcting the beta-hemoglobin mutation responsible for sickle cell anemia and a mutation in the tumor suppressor gene *p53* capable of contributing to many forms of cancer.

Gene therapy can also be achieved through **RNA editing**, a process where changes are made in the sequence of RNA after it is copied from

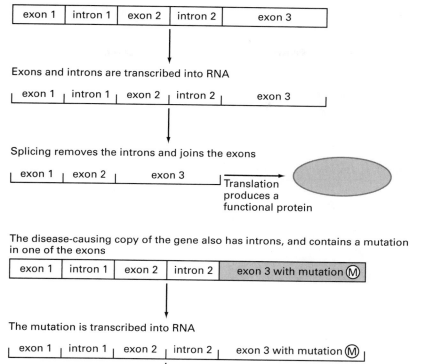

FIGURE 21. Outcomes of RNA splicing in normal cells (top) and those from patients with a disease-causing gene.

DNA. As a result of RNA editing, more than one protein can be made from a single set of instructions in the DNA. This process plays a role in the production of two different forms of apoB, a component of lipid carriers that distribute cholesterol and other fats throughout the body via the bloodstream. Produced in the liver, apoB^{100} contributes to the formation of lipid carriers associated with elevated risk for cardiovascular disease. An RNA editing process changes the signal for one of the amino acids to a stop signal. The result is a shorter version of the protein, apoB^{48}, which is normally produced in the digestive system and does not have the damaging effects of apoB^{100}. Delivering a gene for the enzyme that controls this editing process

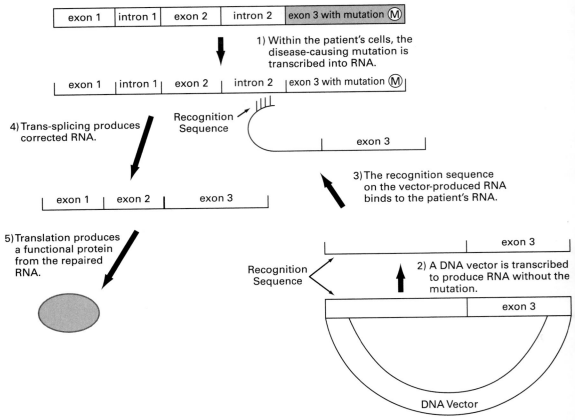

FIGURE 22. Trans-splicing produces a corrected RNA copy of a defective gene.

is a potential treatment for hyperlipidemia, a condition where high levels of fats in the blood can lead to buildup of artery-clogging plaque, contributing to heart and blood vessel disease.

TREATING COMPLEX CONDITIONS RESULTING FROM MULTIPLE GENE DEFECTS

Other exciting developments show that gene therapy is not limited to treating single gene disorders. Cancer, for example, is known to be the result of multiple genetic changes. Rather than trying to reverse each of these mutations individually, scientists have focused on interfering with characteristics of cancer cells that arise from the complex interaction of genetic changes. In one approach, a cancer vaccine is being tested that will stimulate the immune system cells to specifically target cancer cells for destruction. This is directed toward an enzyme called *telomerase,* which restores telomeres, protective structures located at both ends of the linear human chromosomes. Unlike the low levels of telomerase activity in most body cells, the action of this enzyme is highly pronounced in cancer cells. In a clinical trial currently underway, RNA for the telomerase is introduced into the patient's immune system dendritic cells, which can display a portion of the telomerase on their surface as a signal to other cells. When the dendritic cells are returned to the body, they work in concert with cytotoxic T cells to target prostate cancer cells. Another cancer success was achieved for glioma cells when an adenovirus was genetically engineered to enhance its binding to mouse tumor cells and was able to destroy those cells in the infective process.

If the disease-causing cells are not easily accessible, gene therapy can fortify neighboring cells to help prevent some changes associated with the condition. In an attempt to help patients with Alzheimer's disease, genes for Nerve Growth Factor were introduced into human fibroblasts, connective tissue cells obtained from the person's skin. These cells were surgically delivered to the brain where their role was to prevent the death of nerve cells that occurs as the disease progresses. Initial results show increased activity for neural cells and improved memory in these patients.

Conclusion

What lies ahead on the pathway for gene therapy? Will this road, paved with the sacrifices of many patients and their families and the hours of dedication and experimentation of researchers, lead to effective treatments for genetic diseases? While it may take a decade or more to accomplish, there is certainly hope that reliable methods will be found to deliver the genetic instructions that will help defective cells produce their own medications. Imagine the freedom that someone would experience if their suffering could be ended by a prescription that never needs refilling! Such is the promise of gene therapy.

Resources for Students and Educators

FOR STUDENTS

Web Resources

All the Virology on the Net (http://www.virology.net/garryfavwebgenether.html) Links to a variety of gene therapy references.
American Cancer Society (http://www.cancer.org/docroot/ETO/content/ETO_1_3X_Gene_Therapy_Questions_and_Answers.asp) Description of gene therapy for cancer and links to additional resources.
American Society of Gene Therapy (http://www.asgt.org/) News releases on advances in gene therapy.

Canavan Disease as presented by the National Institutes of Health
(http://www.ninds.nih.gov/disorders/canavan/canavan.htm) General information on Canavan disease and links to additional resources.

Clinical Trials as presented by the National Institutes of Health
(http://clinicaltrials.gov) Searchable database of medical studies that are seeking patient participants. Use "gene therapy" as a search term to find treatments that are currently being tested.

Cystic Fibrosis Foundation (http://www.cff.org/about_cf/gene_therapy_and_cf/) Describes gene therapy attempts for cystic fibrosis.

Howard Hughes Medical Institute: Blazing a Genetic Trail (http://www.hhmi.org/genetictrail) Excellent Web site that provides actual stories of gene discovery (such as the search for the cystic fibrosis gene) and dilemmas presented by genetic testing and gene therapy.

Human Genome Project Gene Therapy Information (http://www.ornl.gov/sci/techresources/Human_Genome/medicine/genetherapy.shtml) Comprehensive description of gene therapy along with extensive links to additional information.

Medline Plus presented by the National Library of Medicine and National Institutes of Health (http://www.nlm.nih.gov/medlineplus/genesandgenetherapy.html) Searchable site with news and reliable health-related information.

Microbiology@Leicester (http://www-micro.msb.le.ac.uk/3035/peel/peel1.html) Thorough description of viral vectors used for gene therapy.

National Cancer Institute (http://www.cancer.gov/cancertopics/factsheet/Therapy/gene) Gene therapy for cancer presented in a question and answer format.

National Institutes of Health (http://history.nih.gov/exhibits/genetics/) Online exhibit "A Revolution in Progress: Human Genetics and Medical Research."

National Library of Medicine (http://ghr.nlm.nih.gov/info=gene_therapy/) Safety and ethics of gene therapy.

National Parkinson's Foundation (http://www.parkinson.org/site/pp.asp?c=9dJFJLPwB&b=308861) Explanation of gene therapy for Parkinson's disease.

Online Mendelian Inheritance in Man (http://www.ncbi.nlm.nih.gov/entrez/query.fcgi?db=OMIM) Outstanding site to search for information on human disease genes and progress in gene therapy.

Stanford University, Huntington's Outreach Project for Education (http://www.stanford.edu/group/hopes/treatmts/pbuildup/h2.html) Explanation of the use of RNA interference technology in gene therapy for Huntington's Disease.

Stanford University, Garry Nolan Laboratory (http://www.stanford.edu/group/nolan/tutorials/tutorials.html) Tutorial on retroviral construction.

U.S. Food and Drug Administration (http://www.fda.gov/cber/infosheets/genezn.htm) Describes the role of the FDA in evaluating gene therapy treatments.

University of Iowa, Center for Gene Therapy of Cystic Fibrosis and Other Genetic Diseases (http://genetherapy.genetics.uiowa.edu/) Links to research articles on gene therapy.

Virology Down Under: Gene Therapy Sites (http://www.uq.edu.au/vdu/GeneTherapylinks.htm) News items and headlines related to gene therapy.

Your Genes, Your Choices (http://www.ornl.gov/hgmis/publicat/genechoice/contents.html) Site explores issues raised by genetic research. Contains excellent case situations and ethical dilemmas for student discussion.

Popular Books

Espejo, R. (Ed.). (2004). *Gene therapy.* At Issue series. Farmington Hills: Greenhaven Press.

Naff, C. F. (Ed.). (2004). *Gene therapy.* Exploring Science and Medical Discoveries. Farmington Hills: Greenhaven Press.

Panno, J. (2004). *Gene therapy: Treating disease by repairing genes.* New York: Facts on File.

FOR EDUCATORS

Web Resources

Nova scienceNow: RNAi (http://www.pbs.org/wgbh/nova/sciencenow/3210_02.html) (http://www.pbs.org/wgbh/nova/teachers/programs/3210_02_nsn.html) Video segment detailing the discovery of RNA interference and its potential use in gene therapy along with educator's guide.

Stanford University, Human Gene Therapy Program (http://www.med.stanford.edu/genetherapy/education/lectures.html) PowerPoint presentation on gene therapy by Dr. Mark Kay, Director of the Program in Human Gene Therapy and Professor in the Department of Pediatrics and Genetics at Stanford University School of Medicine.

University of British Columbia (http://bioteach.ubc.ca/TeachingResources/Applications/GeneTherapyPckVanWier.html) Classroom activities related to gene therapy.

University of Utah, Genetic Science Learning Center (http://gslc.genetics.utah.edu/units/genetherapy/index.cfm) "Gene Therapy: Molecular Bandage?" presents treatment of cystic fibrosis as a case study in gene therapy.

Virginia Commonwealth University Life Sciences (http://www.pubinfo.vcu.edu/secretsofthesequence/) Free access to video "Sickle Cell Anemia, Hope from Gene Therapy" and associated class activities.

Journal Articles

Aiuti, A., et al. (2002). Correction of ADA-SCID by stem cell gene therapy combined with nonmyeloablative conditioning. *Science,* 296, 2410–2413.

Byun, J., et al. (2003). Efficient and specific repair of sickle beta-globing RNA by trans-splicing ribozymes. *RNA,* 9(10), 1254–1263.

Fischer, A., & Cavazzana-Calvo, M. (2006). Whither gene therapy? *The Scientist,* 20, 36.

Goyenvalle, A., et al. (2004). Rescue of dystropic muscle through U7 snRNA-mediated exon skipping. *Science, 306*(5702), 1796–1799.

Hacein-Bey-Albina, S., et al. (2002). Sustained correction of X-linked severe combined immunodeficiency by ex vivo gene therapy. *New England Journal of Medicine, 346*(16), 1185–1193.

Hacein-Bey-Albina, S. et al., (2003). LMO2-associated clonal T cell proliferation in two patients after gene therapy for SCID-X1. *Science, 302*: 415-419.

Janson, C., et al. (2002). Gene therapy of Canavan disease: AAV-2 vector for neurosurgical delivery of aspartoacylase gene (ASPA) to the human brain. *Human Gene Therapy, 13*(11), 1391–1412.

Podsakoff, G. M., Engel, B. C., & Kohn, D. B. (2005). Perspectives on gene therapy for immune deficiencies. *Biology of Blood and Marrow Transplantation, 11*(12), 972–976.

Porteus, M. H., & Carroll, D. (2005). Gene targeting using zinc finger nucleases. *Nature Biotechnology, 23*, 967–973.

Shin, K. S., Sullenger, B. A., & Lee, S. W. (2004). Ribozyme-mediated induction of apoptosis in human cancer cells by targeted repair of mutant p53 RNA. *Molecular Therapy, 10*(2), 365–372.

Tuszynski, M. H., et al. (2005). A phase 1 clinical trial of nerve growth gene therapy for Alzheimer disease. *Nature Medicine, 11*(5), 551–555.

Books

Curiel, D. T., & Douglas, J. T. (2002). *Adenoviral vectors for gene therapy.* San Diego: Academic Press.

Factor, P. H. (2001). *Gene therapy for acute and acquired diseases.* Boston: Kluwer Academic Publishers.

Flotte, T. R., & Berns, K. I. (2005). *Adeno-associated viral vectors for gene therapy. Laboratory techniques in biochemistry and molecular biology.* (Vol. 31). San Diego: Elsevier Press.

Kaplitt, M. G., & During, M. J. (Eds.). (2006). *Gene therapy in the central nervous system: From bench to bedside.* San Diego: Elsevier Press.

Lattime, E. C., & Gerson, S. L. (2002). *Gene therapy of cancer: Translational approaches from preclinical studies to clinical implementation* (2nd ed.). San Diego: Academic Press.

Palladino, M. A. (2006). *Understanding the Human Genome Project* (2nd ed.). San Francisco: Benjamin Cummings.

Redberry, G. W. (2005). *Trends in gene therapy research.* New York: Nova Biomedical Books.

Schleef, M. (2005). *DNA pharmaceuticals: Formulation and delivery in gene therapy, DNA vaccination and immunotherapy.* Weinheim: Wiley VCH.

Note: All of the Web sites and links presented in this booklet were last accessed and verified for accuracy on May 15, 2006.